Carol drives a tractor.
Just like her husband, Ron.

They work from dawn til dusk

Whoosh, Whoo

to get the farming done.

Chuk, Chuk, Chuk, Chuk

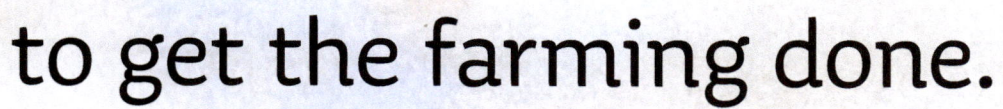

Their farm has many acres.
As far as you can see.

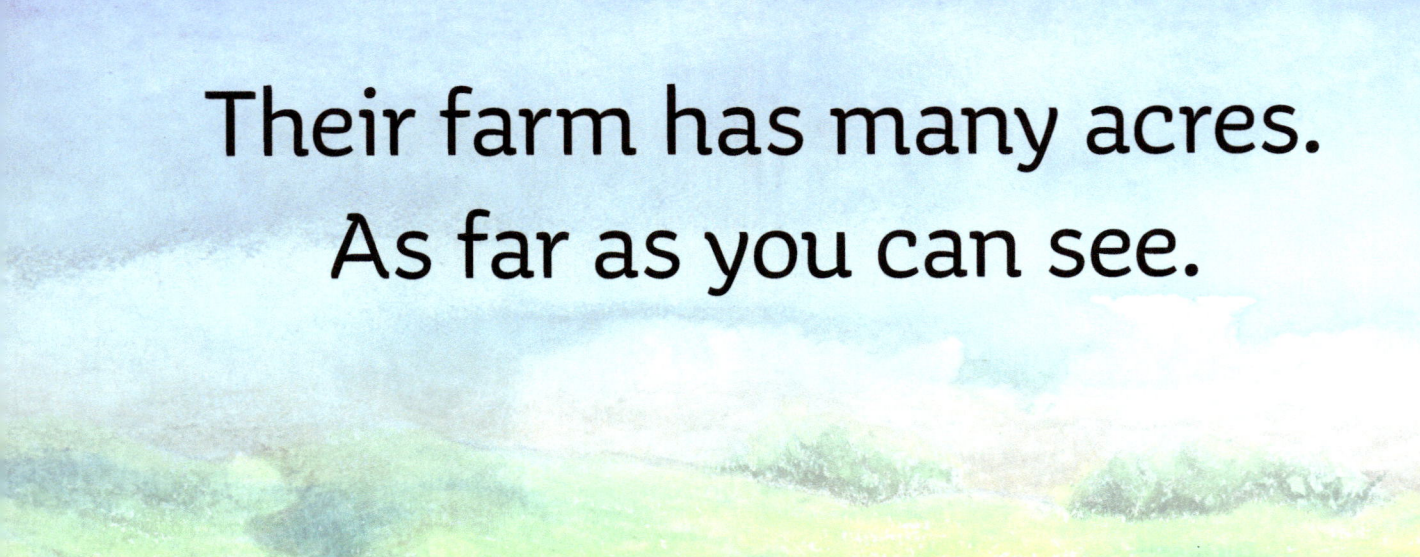

When everything is planted,
it's a sea of waving green.

Carol pulls a grain cart.
It holds a lot of wheat.

Clackety-Clack

With the cultivator
she makes the tree rows neat.

Vrrrrr...   *Vrrrrr...*

At times she drags a roller
to press the big rocks down.

Sometimes she grades the roads,

Crunch Crunch
Crunch

so the bus can get to town.

Of all the tractors made,
she can drive them all!

From zero turn lawn mowers,

to combines
really tall.

Some of them have cabs.

Others don't have mirrors.

But her favorite of all has to be John Deere!

While Curtis drives a swather,

Swish, Swish.   Swish, Swish.   Swish, Swish.

Tammi drives the rake.

And Cale drives a combine,
cuz he's got grain to make!

Wrrrrr... Wrrrrr... Wrrrrr...

Jade drives a baler.
Sometimes right through town!

But when Carol drives a tractor,

She wears a pink ball gown!

Just kidding!

She wears blue jeans.

You can learn to drive a tractor, too! It isn't just for dreams!

www.ingramcontent.com/pod-product-compliance
Lightning Source LLC
Chambersburg PA
CBHW041623120626
46551CB00003B/562